CARBON COMPOUNDS

Preface

This book is genuinely written for grasping the fundamental concept of chemistry. It is aimed to the secondary level students. It can serve as a reference for a particular topic. It is also useful for various competitions.

It precisely deals with carbon compounds, their properties, reactions and uses in daily life. Second chapter deals with the periodic classification of elements.

Introduction

Necessity is the mother of invention. Human curiosity and imagination had added a wing to our flight of invention. Science has no destination but journey. As a domain of science chemistry also deals with what? Why? How? From what the matter around us made up of? And what are the properties of these matters? These are the subjects of chemistry. It has a golden history from alchemists to the detection of higgs boson (commonly referred by journalists "the god particle"). It is assumed that alchemists used to convert other elements into gold. Now we are observing the same thing only the difference between two elements is the number of fundamental particles electron, proton, and neutrons. Now, we are aware of that the whole universe is made up of 118 elements (discovered till now). Our body, the daily items we use, medicine we take, and our surroundings all are the result of these elements only.

Index

Chapter 1

Carbon and its compounds

Introduction

Carbon is present in almost every matter around us. Maximum things around us on burning changed to the ashes (oxides of carbon). For example: the sugar we eat, cloths we wear and food items we eat all produces carbon on burning. It is such a versatile and Omni present element; it forms maximum no. of compounds in the world.

About carbon

As we have already studied its atomic no. is 6 and its atomic mass is 12 amu. It has 6 protons and 6 electrons in a C-12 carbon atom. During reaction whether it can take 4 electrons or loose 4 electrons in order to get a stable configuration. In first case 6 protons has to balance 10 electrons which seems unstable, in second case it requires much more energy to release 4 electrons. To overcome this problem carbon shares it electrons with other elements or same type carbon atoms.

The compounds which are formed from the combination of hydrogen and carbon are generally referred as hydrocarbons. However these compounds can have some other elements like sulphur, oxygen, nitrogen etc as a functional group. About functional group we will study in sub sequent sections.

Covalent bond

We have already studied in part 1, about covalent bonding. Here we will see that in covalent bonding the shared electrons is counted in both atoms to full fill the stable configuration. These bonding also can be expressed in the form of electron dot structure. Each electron is represented as a dot. The combination of two dots forms a single bond.

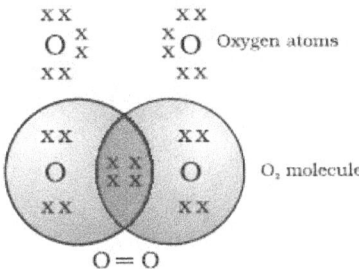

Oxygen atoms

O₂ molecule

$O = O$

Versatile nature of carbon

There are two versatile property of carbon through which it forms infinite no. of carbon compounds. These are as follows:

1. Its four valency which provides a lot of branching.

2. Its property to link with same type of atoms generally known as *catenation*.

Saturated and unsaturated carbon compounds

Further hydrocarbons can be divided in two categories i.e. saturated and unsaturated. Saturated hydrocarbons contain single bonds while unsaturated contain double or triple bonds. There other differences also between them, for example: saturated hydrocarbons give blue flame on complete combustion while unsaturated give a yellow sooty flame. Saturated hydrocarbons undergo substitution reaction while unsaturated can't. Generally saturated are fairly inert in the presence of most of the reagents while unsaturated are not. Alkanes are saturated while alkenes and alkynes are unsaturated in nature.

Alkane

Alkanes have general formula C_nH_{2n+2}, where n refers to the no. of atoms. This formula shows that the no. hydrogen atom is two more than the double of the carbon atom. Meth, eth, prop but,pent, hex, hept, oct, non and dec signifies the no. of carbon atom 1, 2, 3, 4 5, 6, 7, 8, 9 and 10 respectively while –ane is joined to these root words to form the name of the compounds. The following table shows the names of first ten alkanes.

Chemical Formula	IUPAC Name
CH_4	Methane
C_2H_6	Ethane
C_3H_8	Propane
C_4H_{10}	Butane
C_5H_{12}	Pentane
C_6H_{14}	Hexane
C_7H_{16}	Heptane
C_8H_{18}	Octane
C_9H_{20}	Nonane
$C_{10}H_{22}$	Decane

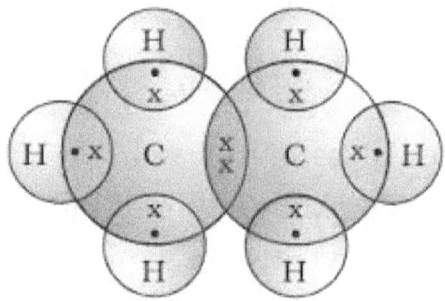

Alkene

Alkenes have general formula C_nH_{2n} , where n refers to the no. of atoms. This formula shows that the no. hydrogen atom is double of the carbon atom. In this compound there must be one double bond.

Alkyne

Alkynes have general formula C_nH_{2n-2}, where n refers to the no. of atoms. This formula shows that the no. hydrogen atom is two less than the double of the carbon atom. In this compound there must be one triple bond.

Butyne	C_4H_6	$$H-C \equiv C - \overset{\overset{\displaystyle H}{\mid}}{\underset{\underset{\displaystyle H}{\mid}}{C}} - \overset{\overset{\displaystyle H}{\mid}}{\underset{\underset{\displaystyle H}{\mid}}{C}} - H$$
Pentyne	C_5H_8	$$H-C \equiv C - \overset{\overset{\displaystyle H}{\mid}}{\underset{\underset{\displaystyle H}{\mid}}{C}} - \overset{\overset{\displaystyle H}{\mid}}{\underset{\underset{\displaystyle H}{\mid}}{C}} - \overset{\overset{\displaystyle H}{\mid}}{\underset{\underset{\displaystyle H}{\mid}}{C}} - H$$

Functional groups

When one hydrogen atom is removed from the alkane, a new group is formed this is known as *alkyl group*. It is generally denoted by –R. Functional group is the atom or group of atoms which when attached with alkyl group can change its physical and chemical properties.

Halo- group

These are the halogen atoms, such as fluorine, chlorine, bromine, iodine etc.

1-chloroethane

$$H-\overset{\overset{\displaystyle H}{\mid}}{\underset{\underset{\displaystyle H}{\mid}}{C}}-\overset{\overset{\displaystyle H}{\mid}}{\underset{\underset{\displaystyle H}{\mid}}{C}}-Cl$$

Alcohol group

This group consists of hydrogen and oxygen i.e. –OH.

```
  H  H
  |  |
H-C--C-O-H
  |  |
  H  H
```

Aldehyde group

Aldehyde group consists of C, H and O in the following fashion:

$$R - \overset{\overset{\displaystyle O}{\|}}{C} - H$$

Ketone group

This group is –CO. in this group there is double bond between carbon and oxygen. And carbon is attached with at least two alkyl group.

$$CH_3 - \overset{\overset{\displaystyle O}{\|}}{C} - CH_3$$

Acetone

Aldehyde and ketone

However this double bond between carbon and oxygen also present in the aldehyde group but it is ketone only in which carbon atom of functional group is attached to at least two alkyl groups.

Carboxylic acid

Carboxylic acid has the functional group –COOH. There is one double bond present between the carbon and oxygen, while there is single bond between hydrogen and oxygen.

Nomenclature of carbon compounds

Naming a carbon compound can be done by the following method –
(i) Identify the number of carbon atoms in the compound. A compound having three carbon atoms would have the name propane.
(ii) In case a functional group is present, it is indicated in the name of the compound with either a prefix or a suffix (as given in Table 4.4).
(iii) If the name of the functional group is to be given as a suffix, the name of the carbon chain is modified by deleting the final 'e' and adding the appropriate suffix. For example, a three-carbon chain with a ketone group would be named in the following manner – Propane – 'e' = propan + 'one' = propanone.
(iv) If the carbon chain is unsaturated, then the final 'ane' in the name of the carbon chain is substituted by 'ene' or 'yne' as given in Table 4.4. For example, a three-carbon chain with a double bond would be called propene and if it has a triple bond, it would be called propyne.

Table 4.4 Nomenclature of functional groups

Functional group	Prefix/Suffix	Example	
1. Halogen	Prefix-chloro, bromo, etc.	$H-\overset{\overset{H}{\mid}}{C}-\overset{\overset{H}{\mid}}{C}-\overset{\overset{H}{\mid}}{C}-Cl$	Chloropropane
		$H-\overset{\overset{H}{\mid}}{C}-\overset{\overset{H}{\mid}}{C}-\overset{\overset{H}{\mid}}{C}-Br$	Bromopropane
2. Alcohol	Suffix - ol	$H-\overset{\overset{H}{\mid}}{C}-\overset{\overset{H}{\mid}}{C}-\overset{\overset{H}{\mid}}{C}-OH$	Propanol
3. Aldehyde	Suffix - al	$H-\overset{\overset{H}{\mid}}{C}-\overset{\overset{H}{\mid}}{C}-C=O$	Propanal
4. Ketone	Suffix - one	$H-\overset{\overset{H}{\mid}}{C}-\overset{\overset{\parallel}{C}}{O}-\overset{\overset{H}{\mid}}{C}-H$	Propanone
5. Carboxylic acid	Suffix - oic acid	$H-\overset{\overset{H}{\mid}}{C}-\overset{\overset{H}{\mid}}{C}-\overset{\overset{O}{\parallel}}{C}-OH$	Propanoic acid
6. Double bond (alkenes)	Suffix - ene	$H-\overset{\overset{H}{\mid}}{C}-\overset{\overset{H}{\mid}}{C}=C\overset{H}{\underset{H}{}}$	Propene
7. Triple bond (alkynes)	Suffix - yne	$H-\overset{\overset{H}{\mid}}{C}-C\equiv C-H$	Propyne

Chemical properties of carbon compounds

Combustion: In combustion reaction the hydrocarbons on combustion produces heat and light.

(i) $C + O_2 \rightarrow CO_2$ + heat and light
(ii) $CH_4 + O2 \rightarrow CO_2 + H_2O$ + heat and light
(iii) $CH_3CH_2OH + O2 \rightarrow CO_2 + H_2O$ + heat and light

Oxidation: oxidation refers to the addition of oxygen. This reaction is generally used to convert alcohols to carboxylic groups. When we add oxygen to alcohol using an oxidizing agent like alkaline $KMnO_4$ or acidified $K_2Cr_2O_7$, alcohol changes to carboxylic acid.

$$CH_3 - CH_2OH \xrightarrow[\text{Or acidified } K_2Cr_2O_7 + \text{Heat}]{\text{Alkaline } KMnO_4 + \text{Heat}} CH_3COOH$$

Addition reaction: addition reaction is used to convert unsaturated hydrocarbon into saturated hydrocarbon. In this reaction a single bond is broken to associate with halogens. In many cases in place of halogen H_2 is used with nickel as a catalyst. it is commercially used to convert vegetable oil into animal fats.

Substitution reaction: Saturated hydrocarbons are fairly unreactive and are inert in the presence of most reagents. However, in the presence of sunlight, chlorine is added to hydrocarbons in a very fast reaction. Chlorine can replace the hydrogen atoms one by one. It is called a substitution reaction because one type of atom or a group of atoms takes the place of another. A number of products are usually formed with the higher homologues of alkanes.

$CH_4 + Cl_2 \rightarrow CH_3Cl + HCl$ (in the presence of sunlight)

Some important carbon compounds

Ethanol

Physical properties

Ethanol is a liquid at room temperature (refer to Table 4.1 for the melting and boiling points of ethanol). Ethanol is commonly called alcohol and is the active ingredient of all alcoholic drinks. In addition, because it is a good solvent, it is also used in medicines such as tincture iodine, cough syrups, and many tonics. Ethanol is also soluble in water in all proportions. Consumption of small quantities of dilute ethanol causes drunkenness. Even though this practice is condemned, it is a socially widespread practice. However, intake of even a small quantity of pure ethanol (called absolute alcohol) can be lethal. Also, long-term consumption of alcohol leads to many health problems.

Chemical properties

(i) Reaction with sodium –
$$2Na + 2CH_3CH_2OH \rightarrow 2CH_3CH_2O\text{–}Na^+ + H_2$$
$$\text{(Sodium ethoxide)}$$
Alcohols react with sodium leading to the evolution of hydrogen. With ethanol, the other product is sodium ethoxide.

(ii) Reaction to give unsaturated hydrocarbon: Heating ethanol at 443 K with excess concentrated sulphuric acid results in the dehydration of ethanol to give ethene –

$$CH_3 CH_2OH \rightarrow CH_2=CH_2 + H_2O$$

The concentrated sulphuric acid can be regarded as a dehydrating agent which removes water from ethanol.

Ethanoic acid

Physical properties

Ethanoic acid is commonly called acetic acid and belongs to a group of acids called carboxylic acids. 5-8% solution of acetic acid in water is called vinegar and is used widely as a preservative in pickles. The melting point of pure ethanoic acid is 290 K and hence it often freezes during winter in cold climates. This gave rise to its name glacial acetic acid. The groups of organic compounds called carboxylic acids are obviously characterized by a special acidity. However, unlike mineral acids like HCl, which are completely ionized, carboxylic acids are weak acids.

Chemical properties

(i) *Esterification reaction:* Esters are most commonly formed by reaction of an acid and an alcohol. Ethanoic acid reacts with absolute ethanol in the presence of an acid catalyst to give an ester –

$$CH_3COOH + CH_3CH_2OH \rightarrow CH_3\text{-}CO\text{-}CH_2CH_3$$

(ethanoic acid)   (alcohol)   acid   (Ester)

Esters are sweet-smelling substances. These are used in making perfumes and as flavouring agents. Esters react in the presence of an acid or a base to give back the alcohol and carboxylic acid. This reaction is known as saponification because it is used in the preparation of soap.

Formation of ester

$$CH_3COOC_2H_5 \, C \, H \, OH \rightarrow C_2H_5OH + CH_3COOH$$
$$(NaOH)$$

(ii) *Reaction with a base:* Like mineral acids, ethanoic acid reacts with a base such as sodium hydroxide to give a salt (sodium ethanoate or commonly called sodium acetate) and water:

$$NaOH + CH_3COOH \rightarrow CH_3COONa + H_2O$$

(iii) *Reaction with carbonates and hydrogen carbonates*: Ethanoic acid reacts with carbonates and hydrogen carbonates to give rise to a salt, carbon dioxide and water. The salt produced is commonly called sodium acetate.

$$2CH_3COOH + Na_2CO_3 \rightarrow 2CH_3COONa + H_2O + CO_2$$
$$CH_3COOH + NaHCO_3 \rightarrow CH_3COONa + H_2O + CO_2$$

Soaps and detergents

Detergents are ammonium or sulphonate salts of long chain carboxylic acids while soaps are sodium salt of carboxylic acid.

When calcium or magnesium salts are dissolved in water it becomes hard water and can't be useful for soap due to formation of scum an insoluble substance. However detergent can be used in hard water because charged ends of detergent do not form scum.

The following table shows the difference between the soaps and detergent.

SOAPS	DETERGENTS
They contain sodium carboxylate (COONa) group.	They contain sodium sulphonate (SO_3Na) group.
They are not suitable for washing with hard water.	They are suitable for both hard and soft water.
They have relatively weak cleansing action.	They have strong cleansing action.
They are biodegradable.	Most of them are non-biodegradable.

Cleaning action of soaps

The action of soaps and detergents is based on the presence of both hydrophobic and hydrophilic groups in the molecule and this helps to emulsify the oily dirt and hence its removal.

When detergent or soaps are dissolved in water they produce many molecules which have hydrophobic and hydrophilic ends. Hydrophobic end repel water molecule and get attached to the dirt while hydrophilic does its opposite. In this process all hydrophobic parts of molecules get attached to the dirt and pulled by water molecules, this unique formation is named *micelle*. When we wash out water dirt is removed and the cloth gets cleaned.

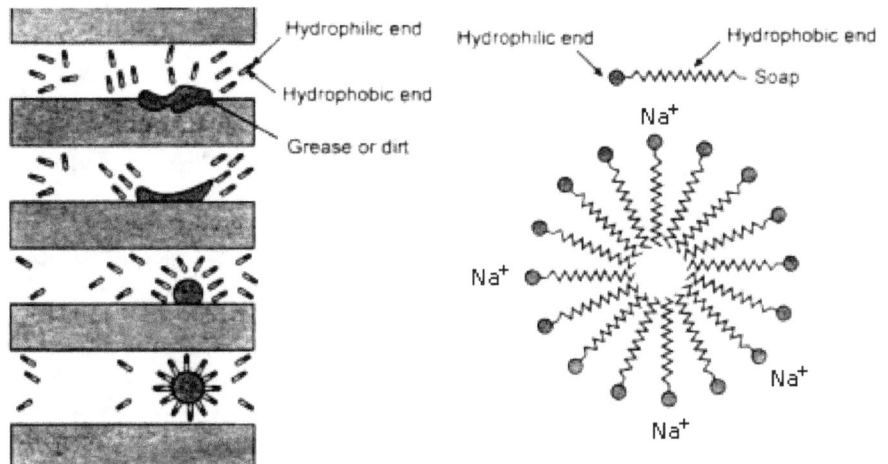

Review questions

OBJECTIVE QUESTIONS:

Q1. The % of carbon in earth crust is?

    a. 0.002

    b. 0.02

    c. 20

    d. 0.2

Q2. Carbon has valency?

    a. 1

    b. 2

    c. 3

    d. 4

Q3. Catenation is found in?

    a. C

    b. O

    c. F

    d. None of these

Q4. Bonding in carbon is?

a. covalent

 b. ionic

c.co-ordinate

d. none of these

Q5. A saturated hydrocarbon has?

    a. Single bond

    b. Double bond

c. Triple bond

d. None of these

Q6. Formula of benzene is?

 a. C_2H_6

 b. C_6H_6

 c. C_5H_6

 d. None of these

Q7. Homologous series differ by?

 a. CH_2 unit

 b. 14 amu by mass

 c. Both a and b

 d. None of these

Q8. $CH_3CH_2CH_2OH$ is?

 a. alcohol

 b. ketone

 c. carboxylic acid

 d. None of these

Q9. In oxidation reaction the oxidising agent is?

 a. Alkaline $KMnO_4$

 b. Acidified $K_2Cr_2O_7$

 c. Both a and b

 d. None of these

Q10. In addition reaction the catalyst used is?

 a. H_2

 b. Acidified $K_2Cr_2O_7$

 c. nickel

 d. None of these

Q11. The conversion of ethanol to ethanoic acid is?

    a. Oxidation reaction

    b. Addition reaction

    c. Substituition reaction

    d. None of these

Q12. Which is commonly known as alcohol?

    a. ethanol

    b. ethanoic acid

    c. methanol

    d. None of these

Q13. Vinegar is formed from?

    a. ethanol

    b. ethanoic acid

    c. methanol

    d. None of these

Q14. In esterification the catalyst used is?

    a. base

    b. acid

    c. salt

    d. None of these

Q15. Saponification is the reverse process of?

    a. carbocation

    b. esterification

    c. carbocation

    d. None of these

Q16. Ionic end of soap dissolves in?

    a. water

b. oil

c. acid

d. base

Q17. Shampoos are?

    a. soap

    b. detergent

    c. salts

    d. acids

Q.18 which of the following can go under addition reaction?

    a. C_2H_6

    b. C_6H_6

    c. C_5H_6

    d. None of these

Q19. CH_3COOCH_3 is?

    a. soap

    b. acid

    c. base

    d. ester

Q20. Which of the following causes hardness of water?

    a. Calcium salt

    b. Magnesium salt

    c. Both a and b

    d. None of these

Match the following:

Q21.

A	B
C-60	saturated
ethene	allotrophes
methane	unsaturated
carbon	catenation

Q22.

A	B
-OH	carboxyllic acid
-CHO	ketone
-CO-	aldehyde
-COOH	alcohol

Q23.

A	B
nickel	substituition reaction
sun light	addition reaction
alkaline $KMnO_4$	oxidation reaction
hot conc. H_2SO_4	formation of ethene

Short answer type questions:

Q24. What are the versatile properties of carbon?

Q25. Write down differences between saturated and unsaturated hydrocarbons?

Q26. What is homologous series?

Q27. What is addition reaction?

Q28. What is substitution reaction?

Q29. Write down physical properties of ethanol?

Q30. Write down physical properties of ethanoic acid?

Q31. Write down chemical properties of ethanol?

Q32. Write down chemical properties of ethanoic acid?

Long answer type questions:

Q33. Draw the structures for the following compounds.
    (i) Ethanoic acid   (ii) Bromopentane*
    (iii) Butanone   (iv Hexanal.
    *Are structural isomers possible for bromopentane?

Q34. Explain the cleaning action of soap?

Q35. Explain the phenomenon of esterification and saponification in detail with examples.

Chapter 2

Periodic classification of elements

Introduction

Why classification is important?

Classification provides us a easy way to understand the properties and access to the particular element. We can take a simple example; in a mall the items pertaining to same purposes are kept one side. If all items present in the mall are spread and mixed, what will happen? It will create a huge disturbance to customer as well as managing team. If it is classified, it is quite easy for selling as well as purchasing.

If all elements are arranged according to some particular properties, it will be quite easier to select a particular element for a particular purpose. Suppose any engineer wants to form the body part of an aero plane. Then first, he will be attempted to ignore liquids and gases. Then he will try to choose light metals which have high tensile strength. Further he may be interested in other properties too. All these comparisons can be done only when all elements are arranged in a particular way. But before 200 years when classification starts, it was a great quest, how to start it? There came thousands of ways and tables some were discarded; some were improved and gave a platform to next one. In this chapter we will see only some of the classification which proved to the base for modern periodic classification of elements.

In this chapter our strategy will be to have a look at the basic rule for a particular classification then its merits and demerits. We will also deal with modern periodic table in details.

Dobereiner's triads

In the year 1817, Johann Wolfgang Döbereiner, a German chemist, arranged elements in the group of three named them as triads. According to dobereiner

When these three elements are arranged in increasing order of their atomic weight the middle weight is average of remaining two. Following table shows the examples.

Elements	Atomic Mass	Arthmetic Mean
Lithium Sodium Potassium	7 23 39	$\dfrac{7+39}{2}=23$
Chlorine Bromine Iodine	35.5 80 126.5	$\dfrac{35.5+126.5}{2}=81$
Calcium Strontium Barium	40 87 137	$\dfrac{137+40}{2}=88$

Demerits

Dobereiner could arrange only three groups.

Newland's octaves

Doberiener's triads encouraged many researchers to search periodicity in elements according to the increasing order of mass. Newlands took the hydrogen as lightest element and thorium as 56th and last element. On arranging these elements in increasing order of their atomic mass, Newlands found that every eighth element shows similar properties. He proposed a word "the octaves law" for this periodicity.

Demerits of Newlands octaves

1. Newlands law of octaves was applicable to only light elements. It was well up to calcium element.
2. Newland announces that there are only 56 elements in the nature and no further elements will be discovered. But soon it was realized that it is not true.
3. Some elements were put in same slot like cobalt and nickel and both of these were in the column of fluorine which has different properties. Iron which resembles properties same as cobalt was kept far away from it.

Newlands' Octaves

H	Li	Be	B	C	N	O
F	Na	Mg	Al	Si	P	S
Cl	K	Ca	Cr	Ti	Mn	Fe
Co, Ni	Cu	Zn	Y	In	As	Se
Br	Rb	Sr	Ce, La	Zr	Di, Mo	Ro, Ru
Pd	Ag	Cd	U	Sn	Sb	I
Te	Cs	Ba, V	Ta	W	Nb	Au
Pt, Ir	Os	Hg	Tl	Pb	Bi	Th

Mendeleev's periodic table

When Mendeleev started preparing the periodic table there were 63 elements. Mendeleev investigated the periodic function of masses as well as physical and chemical properties. He formed oxides and hydrides of each element and kept similar in a same group. Then he stated the law of periodic table which states that which states that **'the properties of elements are the periodic function of their atomic masses'**.

Achievements of Mendeleev's periodic table:

Mendeleev's left some gaps in the periodic table. He took it as a merit and explained that some elements will be discovered in the future and will be kept at these positions. These elements were scandium, gallium, germanium and were respectively named *as **Eka–boron, Eka–aluminium and Eka–silicon***. Here eka (a sanskrit word) means one. Eka word was prefix to the element preceding in the same group. Predictions of properties were almost same as these elements have.

His periodic table was so flexible that when inert gases were discovered they got their places easily.

Limitations of Mendeleev's periodic table:

Mendeleev was ***unable to give a fix position for hydrogen***. His periodic table was based on physical and chemical properties and hydrogen has these properties similar to metal as well as non metals.

As periodic table was arranged in according to the increasing order of their atomic mass there was ***no any place for isotopes***.

	H 1.01									
He 4.00	Li 6.94	Be 9.01	B 10.8	C 12.0	N 14.0	O 16.0	F 19.0			
Ne 20.2	Na 23.0	Mg 24.3	Al 27.0	Si 28.1	P 31.0	S 32.1	Cl 35.5			
Ar 40.0	K 39.1	Ca 40.1	Sc 45.0	Ti 47.9	V 50.9	Cr 52.0	Mn 54.9	Fe 55.9	Co 58.9	Ni 58.7
	Cu 63.5	Zn 65.4	Ga 69.7	Ge 72.6	As 74.9	Se 79.0	Br 79.9			
Kr 83.8	Rb 85.5	Sr 87.6	Y 88.9	Zr 91.2	Nb 92.9	Mo 95.9	Tc (99)	Ru 101	Rh 103	Pd 106
	Ag 108	Cd 112	In 115	Sn 119	Sb 122	Te 128	I 127			
Xe 131	Ce 133	Ba 137	La 139	Hf 179	Ta 181	W 184	Re 180	Os 194	Ir 192	Pt 195
	Au 197	Hg 201	Ti 204	Pb 207	Bi 209	Po (210)	At (210)			
Rn (222)	Fr (223)	Ra (226)	Ac (227)	Th 232	Pa (231)	U 238				

The modern periodic table

This periodic table was prepared by Henry Moseley in 1913. In this table Moseley took the atomic no. as the as a fundamental property rather than atomic mass. This concept solved many limitations of Mendeleev's periodic table. This law can be stated as follows:

'Properties of elements are a periodic function of their atomic number.'

Position of elements in modern periodic table

In modern periodic table, there are 7 horizontal rows known as periods and 18 vertical columns known as groups.

The elements having same chemical and physical properties are arranged in the same group. However in periods a gradation or slight change in physical and chemical properties is observed.

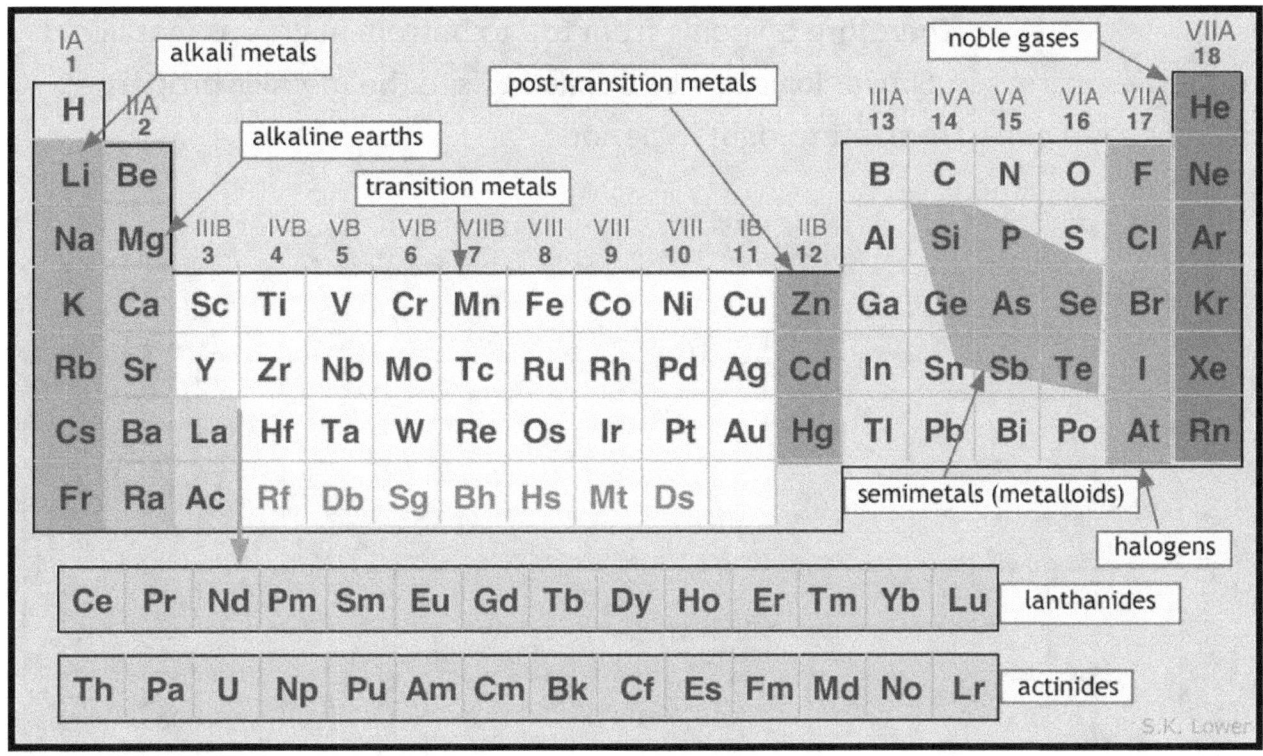

Trends in the modern periodic table

Valency: in a group the valency remains unchanged as there are same no. of electrons in outer most shell of atom. When we move from left to right in a period the valency first increases and then decreases. We have already seen how to calculate valency in "basic chemistry 1" if we take the second period of modern periodic table the elements Li, Be, B, C, N, O, F, and Ne we get the valencies as 1, 2, 3, 4, 3, 2, 1, 0 respectively.

Atomic size: in group as we go from top to bottom, due to increase in no. of orbital atomic size increases. When we move in periods from left to right the no. of orbitals remain same but no. of protons increases which attracts electrons and size shrinks i.e. atomic size decreases.

Metallic and non metallic properties: as we have seen in above section that when we go from top to bottom in groups the atomic size Increases and hence the electrons in the outer most shells are more free to move, this contribute to metallic properties.

Therefore on going from top to bottom in groups metallic properties increases. Similar logic can be applied to see the metallic properties decreases on going from left to right in periods.

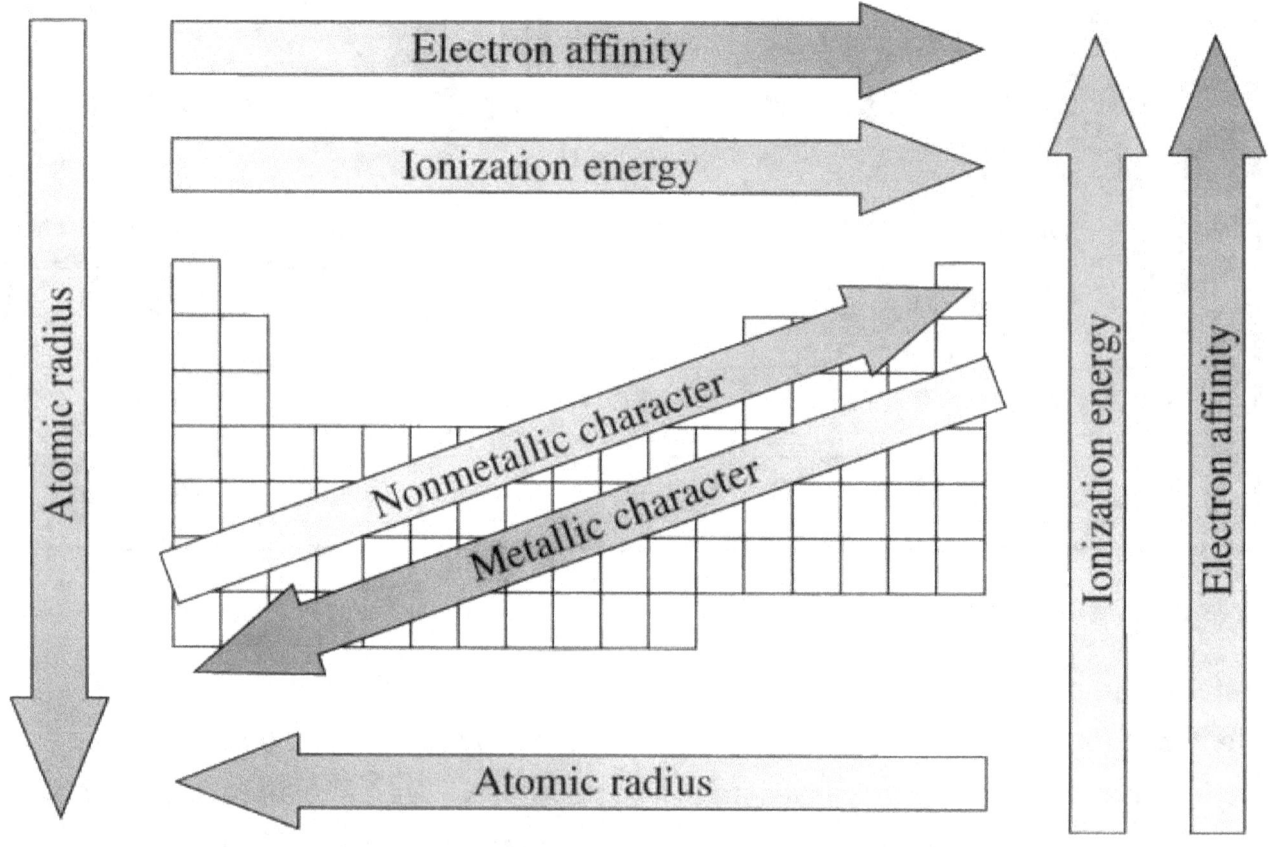

Review questions

Objective questions:

Q1. Doberiener arranged the elements in the form of?

    a. triads
    b. octaves
    c. tetrads
    d. none

Q2. Newlands arranged how many elements?

    a. 56
    b. 63
    c. 114
    d. 118

Q3. Mendeleev left gape for which element in his periodic table?

    a. scandium
    b. gallium
    c. germenium
    d. all of these

Q4. Mendeleev used basic concept of?

    a. Physical properties
    b. Chemical properties
    c. Atomic masses
    d. all of these

Q5. Who prepared modern periodic table?

    a. newland
    b. doberiener
    c. mendeleev
    d. moselley

Q6. How many periods are there in modern periodic table?

    a. 18
    b. 7
    c. 20
    d. None of these

Q7. What is the valency of magnesium?

    a. 1
    b. 2
    c. 3
    d. 4

Q8. How many groups are there in modern periodic table?

    a. 18
    b. 7
    c. 20
    d. None of these

Q9. In periods going right from left, the valency?

    a. increases
    b. decreases
    c. first increases then decreases
    d. None of these

Q10. In groups going from top to bottom, the valency?

    a. increases
    b. decreases
    c. first increases then decreases
    d. remain same.

Q11. In periods going right from left, atomic size?

    a. increases

b. decreases

c. first increases then decreases

d. None of these

Q12. In groups going from top to bottom, the atomic size?

a. increases

b. decreases

c. first increases then decreases

d. remain same.

Q13. In groups going from top to bottom, the metallic properties?

a. increases

b. decreases

c. first decreases then increases

d. remain same.

Q14. In periods going right from left, the non metallic properties?

a. increases

b. decreases

c. first increases then decreases

d. None of these

Match the followings

Q15.

A	B
Law of octaves	moselley
triads	mendeleev
gaps in table	newlands
place for isotopes	doberiener

Q16.

A	B
56 elements	moselley
63 elements	mendeleev
9 elements	newlands
118 elements	doberiener

Long answers type questions:

Q17. Explain the trends in the periodic table for

   a. Valency
   b. Atomic size
   c. Metallic and non metallic properties

Q18. Which element has
(a) Two shells, both of which are completely filled with electrons?
(b) The electronic configuration 2, 8, 2?
(c) A total of three shells, with four electrons in its valence shell?
(d) A total of two shells, with three electrons in its valence shell?
(e) twice as many electrons in its second shell as in its first shell?

Q19. (a) What property do all elements in the same column of the Periodic Table as boron have in common?
(b) What property does all elements in the same column of the Periodic Table as Fluorine has in common?

Q20. The position of three elements A, B and C in the Periodic Table are shown below –

Group 16	Group 17
-	-
-	A
-	-
B	C

(a) State whether A is a metal or non-metal.

(b) State whether C is more reactive or less reactive than A.

(c) Will C be larger or smaller in size than B?

(d) Which type of ion, cation or anion, will be formed by element A?

Chapter 1

Carbon Compounds

Q1.b

Q2.d

Q3.a

Q4.a

Q5.a

Q6.b

Q7.c

Q8.a

Q9.c

Q10.c

Q11.a

Q12.a

Q13.b

Q14.b

Q15.b

Q16.a

Q17.b

Q18.a

Q19.d

Q20.c

Q21.

a---b

b---c

c---a

d---d

Q22.

a---d

b---c

c---b

d---a

Q23.

a---b

b---a

c---c

d---d

Chapter 2

Periodic Classification Of Elements

Q1.a

Q2.a

Q3.d

Q4.d

Q5.d

Q6.b

Q7.b

Q8.a

Q9.c

Q10.d

Q11.b

Q12.a

Q13.a

Q14.a

Q15.

a---c

b---d

c---b

d---a

Q16.

a---c

b---b

c---d

d---a

NOTES